Whale, Dolphin, and Other Pods

Cynthia O'Brien

A Crabtree Crown Book

Crabtree Publishing

crabtreebooks.com

Author: Cynthia O'Brien

Series research and development:
Reagan Miller, Janine Deschenes

Editor: Janine Deschenes

Proofreader: Kathy Middleton

Design and photo research:
Katherine Berti

Print and production coordinator:
Katherine Berti

Image credits:
Istockphoto
Dougall_Photography : p. 24–25 (bottom)
Wikimedia Commons: p. 17 (top right)
All other images from Shutterstock

Crabtree Publishing

crabtreebooks.com 800-387-7650

In Canada: We acknowledge the financial support of the Government of Canada through the Canada Book Fund for our publishing activities.

Hardcover 978-1-0398-0656-6
Paperback 978-1-0398-0682-5
Ebook (pdf) 978-1-0398-0707-5
Epub 978-1-0398-0734-1

Published in Canada
Crabtree Publishing
616 Welland Avenue
St. Catharines, Ontario
L2M 5V6

Published in the United States
Crabtree Publishing
347 Fifth Avenue
Suite 1402-145
New York, New York, 10016

Library and Archives Canada Cataloguing in Publication
Available at Library and Archives Canada

Library of Congress Cataloging-in-Publication Data
Available at the Library of Congress

Printed in the U.S.A./012023/CG20220815

Contents

4 What Is a Pod?

6 Why Do Animals Live in Pods?

8 Whales

10 Humpback Whales

12 Sperm Whales

14 Beluga Whales

16 Narwhals

18 Dolphins

20 Bottlenose Dolphins

22 Orcas

24 Porpoises

26 Hippos

28 Sea Otters

30 Glossary

31 Learning More

32 Index

32 About the Author

What Is a Pod?

People in families or communities live together or in neighborhoods. Have you noticed that many animals also live, hunt, or travel together in groups? Animal groups have different names. Fish swim together in groups called schools. Deer live together in herds. A pod is the name for animals that live, travel, or hunt together in the water. Whales, dolphins, hippos, and sea otters are types of animal **species** that live in pods.

Whales and most dolphins are marine ***mammals****. They live in pods in the world's salty oceans and seas.*

Pods come in different sizes. A pod may include just two animals. Other pods may have 20 or more members. Sometimes pods join together to create even bigger pods. Pods also change in size as babies are born. Adult animals often leave to join or form a new pod.

Some pod animals, such as hippos and river dolphins, live in freshwater.

Why Do Animals Live in Pods?

Living in pods helps animals **breed**, find food, and build social bonds with each other. It also helps to keep them safer from **predators**.

Baby animals in pods depend on their mothers. They also rely on adult members of the pod to keep them safe. Sometimes whale and dolphin mothers form nursery pods with other mothers and their calves. The mothers help one another look after their young.

Humpback whales communicate with other pod members by making sounds that travel underwater. They leap out of the water and splash on its surface. They also slap their fins and tails on the water's surface.

Some pods are made up of family groups. Many other pods have animals that are not related. Bachelor pods, for example, are formed when groups of young male whales join together. Small pods often join other pods to **migrate**. Older animals help to protect the young on the journey.

When many dolphin pods join together, it is known as a "superpod." These can have hundreds or even thousands of dolphins!

Whales

At a Glance

Number of species: More than 80
Pod size: 2 to 30 or more
Location: Oceans worldwide

Due to their size and power, blue whales have almost no predators. They do not need the protection of a pod.

Whale species fall into two main types: **baleen** whales and toothed whales. Baleen whales are usually larger than toothed whales. The blue whale is a baleen. It is the largest animal on Earth and spends most of its time alone. Many other whales belong to pods some or all the time.

At least 70 whale species are toothed, including dolphins and sperm whales. Small toothed whales, such as belugas, often live in larger pods. This gives them protection from large predators, such as polar bears. Many whales also live together because they are very social. They communicate and play with each other.

Did You Know?

Toothed whales use **echolocation** to look for food and find their way in the ocean. First, the whales make sounds. Then they listen for the sounds to echo, or bounce back, from objects in the sea.

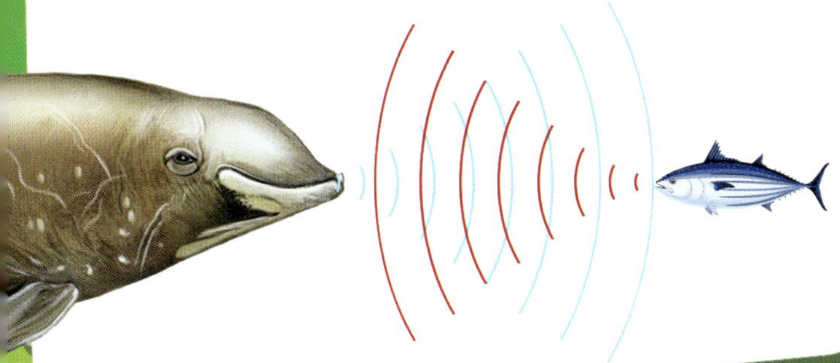

A baleen whale has a unique feeding system inside its mouth. First, it opens its mouth and takes in seawater. Then it pushes the water out. As the water flows out, baleen plates inside its mouth trap any small ***prey*** *in the water.*

The orca is both a type of dolphin and a toothed whale.

Humpback Whales

At a Glance

Number of subspecies: Three—North Atlantic, North Pacific, and Southern **Hemisphere**
Pod size: 2 to 15
Location: Coastal waters of all oceans

Humpback whale songs can be heard under the water from many miles away.

Humpback whales are baleen whales. They migrate long distances from feeding grounds in cold waters to breed in warmer waters. They migrate alone or in small pods. When they arrive, they form larger pods.

Male humpbacks also form competition pods for short times. Several males try to attract one female. They shove each other, slap the water with their **flukes**, or lunge with their heads. The strongest whale wins the female and stays with her.

The males also sing songs, probably to attract females. Each pod has its own song, which changes over time. Sometimes whales share their songs with nearby pods.

Humpbacks sometimes slap their tails on the water's surface before diving down to hunt. This is called lobtail feeding. The whales learn this strategy from each other.

Sticking Together

Humpbacks hunt together. They form a circle and dive deep into the water. As they swim up, they spin and blow bubbles. The bubbles create a net that forces many fish into the center of the circle. Then the whales can open wide and take in their food.

Sperm Whales

At a Glance

Number of subspecies: Three: common sperm whale, pygmy, and dwarf
Pod size: 15 to 20
Location: Mild to warm ocean waters worldwide

The deep-diving sperm whale is the largest toothed whale. These whales often gather in large numbers for short periods, especially during breeding season. Females form pods with other females and their young in their family. They live in these pods their whole lives. While one or more mothers hunt, the other whales look after the calves.

Male sperm whales leave their family pods when they are between four and 21 years old. They may travel alone or form bachelor pods.

Female sperm whales use specific codas with their young.

Sperm whales use groups of clicking sounds to communicate. These are called codas. Each coda has a different purpose. A coda may mean a whale is tracking prey or warn that a whaling ship is nearby. Sperm whales also have individual codas. These different codas help the whales recognize each other.

Sperm whales dive up to 3,280 feet (1,000 m) under the surface. They can stay underwater for over an hour.

Sticking Together

Sperm whales work together to keep away attackers, such as orcas. With their tails out, the whales form a circle around their young. Together, the whales slap the water forcefully with their tails. This makes it impossible for orcas to come near.

Beluga Whales

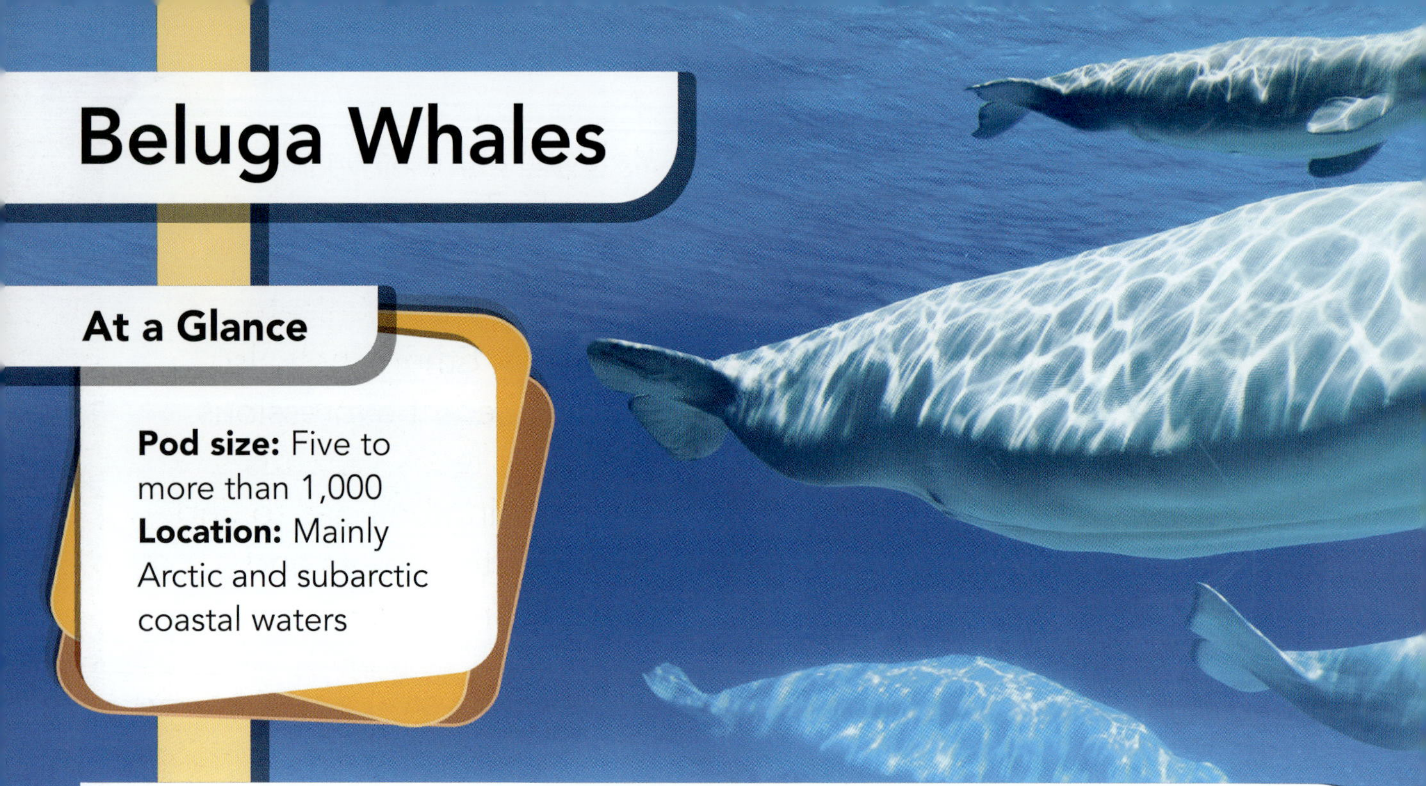

At a Glance

Pod size: Five to more than 1,000
Location: Mainly Arctic and subarctic coastal waters

Belugas are small, toothed whales. They live in pods for protection, to hunt, and to **socialize**. The largest adult male usually leads the pod. Many pods often come together to migrate. This keeps the belugas safer from predators. Females also form large nursery pods with other cows and calves. They teach their young how to hunt and survive. Belugas also move between pods. They make friends with new groups.

Baby belugas are a gray color. They slowly lose their color and turn white around age seven to nine.

Belugas are noisy and sociable. They make chirping, whistling, and clicking sounds to communicate with each other. They also make different expressions with their faces. Scientists are studying belugas to find out what the noises and expressions mean.

The top of a beluga's head is a squishy mass called a melon. The melon moves when the beluga makes sounds.

Most of the time, belugas blow bubbles through their blowholes. They can also use their mouths to blow bubbles.

Did You Know?

Belugas show how they are feeling by blowing bubbles. Small bubbles or a bubble ring usually means belugas are happy and playing. A sudden burst of bubbles may mean a beluga is scared or wants to warn others of danger.

Narwhals

At a Glance

Pod size: 15 to 20
Location: Arctic coastal waters and rivers

The narwhal is closely related to the beluga. It is famous for its long, spiral tusk. Usually, only males have tusks. The tusk is really one of the narwhal's two teeth. Scientists are learning more about it. The tusk may help narwhals hunt. Narwhals sometimes seem to touch tusks in play or affection. Pods of male narwhals also seem to use their tusks in displays to attract mates.

Narwhals move and turn their tusks to look fierce to predators. But they do not use them to fight each other.

In winter, narwhals stay in smaller pods. In summer, hundreds or even thousands of narwhals migrate north to breeding grounds. The pods join to create large groups there. Larger groups bring together males and females for breeding. They also give calves more protection.

Some small narwhal pods have all males. Some are made up of females and their young. Others are mixed pods of males and females.

Did You Know?

Narwhals can live to be more than 125 years old. Their skin changes as they age. Newborn calves are pale gray or brown. Over time, they become marked with black and white spots. Older narwhals turn pale again.

Dolphins

At a Glance

Number of species: Almost 40 ocean dolphins and six river dolphins
Pod size: Five to over 1,000
Location: Ocean dolphins: oceans and seas worldwide; River dolphins: Asia and South America

Dolphins are small, intelligent toothed whales. They gather in pods to feed, teach their young, and stay safe. Dolphins also help other pod members when they are injured or sick.

Large dolphin pods may gather to keep predators away. Adult males watch for danger. If hunting orcas come close, the male dolphins alert the rest of the pod. Right away, the dolphins move very close to each other and swim away as a group.

Female dolphins and their calves also gather in nursery pods. Some females take care of the calves while others hunt for food.

In nursery pods, dolphin calves learn the skills they need to survive from their mothers.

Dolphins like to play in the water together. They jump over waves and race after boats.

Amazon River dolphins are the largest river dolphin. They are known for their unique pink color.

Did You Know?

Dolphins have very good memories and form close friendships with other dolphins. A dolphin can recognize another dolphin's specific sounds, even after not seeing it for many years.

Bottlenose Dolphins

At a Glance

Number of subspecies: Two—common bottlenose and Indo-Pacific bottlenose
Pod size: Two to 15
Location: Warm oceans worldwide

Bottlenose dolphins are noisy and social in their pods. They use different sounds to communicate. Each bottlenose dolphin creates its own whistle not long after it is born. This unique whistle acts like a name. Dolphins respond to their whistle and can be identified by it.

Bottlenose dolphins can swim up to 22 miles per hour (35 kph).

Bottlenose dolphin pods change often as some animals leave and others join. Male dolphins, for example, sometimes form small pods with their best friends.

Dolphins in a pod protect each other. One way is by keeping one eye open while sleeping. They watch for predators and warn the rest of the pod if danger is near.

Bottlenose dolphin pods circle tightly around a group of fish. Then the dolphins take turns swimming through the circle to feed.

Sticking Together

Bottlenose dolphins cooperate to hunt. They teach each other hunting tricks. Some use their tails to smack and stun fish. Other dolphins learn to use a shell or make a ring of mud to trap fish.

Sometimes bottlenose dolphins join other pods of toothed whales. This gives more protection from predators. It can also make hunting easier.

Orcas

At a Glance

Number of subspecies: Multiple **ecotypes**, based on where they live in the world
Pod size: Two to 40
Location: Worldwide in coastal seas and open oceans

Orcas communicate with each other using underwater clicks, whistles, and calls. Different pods have unique calls.

Orcas are the largest dolphins. Two **resident** pods live in the Pacific Ocean near the coasts of British Columbia, Canada, and Washington, U.S.A. Other pods move around the oceans.

An orca stays with its mother's family its whole life. Sons breed with orcas of another family but return to their mother's pod. Older female orcas lead family pods. They teach the young how to hunt and avoid boats.

Family pods spend time with related pods. This creates a clan that understands each other's calls. Several clans form a community. Communities in different parts of the world have their own sounds and ways of hunting.

Norwegian orca pods developed carousel feeding. They surround a school of fish and squeeze it into a tight ball. Then, they slap their tails into the ball of fish, killing many at once.

Sticking Together

Orca pods work together to hunt seals. The orcas swim in a line toward the seals, where they sit on ice. As the orcas get closer, they smack their tails quickly at the same time. This creates a large wave that washes the seals off the ice and into the water.

Orcas often hold their heads above the water's surface to look around. This is called spy-hopping.

Porpoises

At a Glance

Number of species: Seven, including the harbor porpoise
Pod size: Two to 10
Location: Coastal oceans worldwide, freshwater rivers in Asia

Scientists and researchers are still learning more about porpoises. Because they tend to keep to themselves and travel in small groups, it is harder to study them.

Porpoises are related to dolphins. Most live in small pods. A pod may include just a mother and her calf. Porpoises may loosely gather into larger groups when hunting or breeding. They spend most of their time hunting for food, such as fish and squid.

A new study shows that porpoises may hunt as a team. Each animal seems to have a role to play. While some porpoises wait, others swim around a school of fish. Then one porpoise splits the group of fish. The waiting porpoises can then attack as fish break away from the group.

A porpoise mother nurses its calf for about eight months. The calf then switches to fish.

Did You Know?

Dall's porpoise is a type of porpoise that lives in the northern Pacific Ocean, near Alaska. Like some dolphins, these porpoises like to go bow-riding. They ride the waves created by the bows, or fronts, of boats. Some researchers think that bow-riding helps them travel from place to place **efficiently**. Others believe that bow-riding is just for fun.

Hippos

At a Glance

Number of species: Two—common hippo and pygmy hippo
Pod size: 10 to 30
Location: Lakes and rivers in eastern and southeastern Africa; western Africa

Though pygmy hippos live alone, common hippos live together in pods. Hippo groups are also called bloats. Hippos are social animals, but sometimes the groups can get crowded. Over 100 hippos might want space in the water to cool down. This can cause problems in the pods. Large and older males will fight to prove their positions in the group.

Mother hippos are very protective of their calves. The calves stay with their mothers for years.

Hippos are one of the biggest land animals, but predators still attack them. Young hippos are especially **vulnerable** to predators such as lions. Living in a pod helps keep them safe. The dominant male helps protect the pod. He warns predators away by opening his mouth wide, grunting, and showing his teeth.

A hippo spends most of its day in the water. This helps it stay cool. In the evening, it climbs onto land to feed on grasses.

Did You Know?

Hippos release an oily red or pink liquid on their skin. It looks like they are sweating blood! In fact, the liquid helps to keep their skin moist and protect it from the Sun.

Hippos do not swim. Instead, they walk along the bottom of the river. They also bounce off the bottom to move themselves forward.

Sea Otters

At a Glance

Number of species: Three—Asian, northern, and southern
Pod size: Average of 10, but can be much larger at times
Location: Coastal waters of the Pacific Ocean

Sea otters spend most of their lives in the ocean. These small animals can live alone, but they also gather in pods. Sea otter pods are also called rafts. Males and females live in separate pods. Male pods are usually larger. Mothers and their babies, called pups, sometimes form nursery pods. The mothers help protect pups from predators, such as eagles.

Sea otters usually look for food alone, but pods offer protection from predators. This is especially important when sea otters sleep. Sea otters also help each other. For example, sea otters may try to help another sea otter trapped in a net.

The sea otter is the smallest marine mammal in North America. Most live off the coast of Alaska.

Sticking Together

Sea otters often sleep together in groups. The sea otters lie close together on their backs. They wrap themselves in kelp, a type of seaweed, to keep from floating into the ocean.

A mother sea otter keeps her pups close, often holding them on her stomach. When she dives down for food, she wraps her pups in kelp to keep them safe.

Sleeping sea otters also hold hands to keep each other from floating away.

Glossary

baleen A substance that grows on the inside of a baleen whale's mouth, making rows of thin plates that filter food out of seawater

breed To have babies

echolocation An animal's ability to locate objects in its environment by sending out and receiving sounds

ecotypes Distinct groups within a species that are adapted to their specific environments

efficiently Completing a task or activity in a way that works well and does not waste time or energy

flukes A whale's tail fins

Hemisphere Half of Earth, divided along the equator into the Northern and Southern Hemispheres

mammals Warm-blooded animals with backbones. Mammal babies drink milk from their mothers' bodies.

migrate Of animals, to move from one place to another, often according to the season

predators Animals that hunt other animals for food

prey Animals that are hunted by other animals for food

resident Staying in one place for a long period of time

socialize Living and interacting with others in a group

species A group of living things with similar characteristics, in which two animals can have babies

subspecies Two or more distinct categories of one species

vulnerable Able to be easily hurt

Learning More

On the Web

https://ocean.si.edu/ocean-life/marine-mammals/why-do-whales-sing
Check out this video that explains how whales sing to communicate with one another.

https://kids.nationalgeographic.com/animals/mammals/facts/orca
How long are orca teeth? What hunting techniques do orca pods use? Learn the answers to these questions and more.

www.natgeokids.com/uk/discover/animals/sea-life/dolphins
Discover awesome facts about bottlenose dolphins.

Books

Coleman, Clara. *Animal Teamwork: Whales Work Together*. PowerKids Press, 2017.

Spence, Kelly. *Bringing Back the Humpack Whale*. Crabtree Publishing, 2019.

Swanson, Jennifer. *Absolute Expert: Dolphins*. National Geographic Kids, 2018.

Index

Africa 26

Arctic 14, 16

Asia 18, 24, 28

bachelor pods 7, 12, 16, 17, 28

bow-riding 25

breeding 6, 10, 12, 17, 22, 24

carousel feeding 23

codas 13

communication 6, 8, 10, 13, 15, 19, 20, 22

echolocation 9

family pods 7, 12, 22

hunting 4, 11, 12, 14, 16, 18, 21, 22, 23, 24

lobtail feeding 11

migration 7, 10, 14, 17

North America 28

nursery pods 6, 14, 18, 28

oceans 4, 8, 9, 10, 12, 14, 16, 18, 20, 22, 24, 25, 28, 29

predators 6, 8, 14, 16, 18, 21, 27, 28

prey 9, 13

rivers 5, 16, 18, 19, 24, 26, 27

spy-hopping 23

About the Author

Cynthia O'Brien has written many books for children, including books about animals and the natural world. Researching this book, she learned a lot about how different animals live together and help each other.